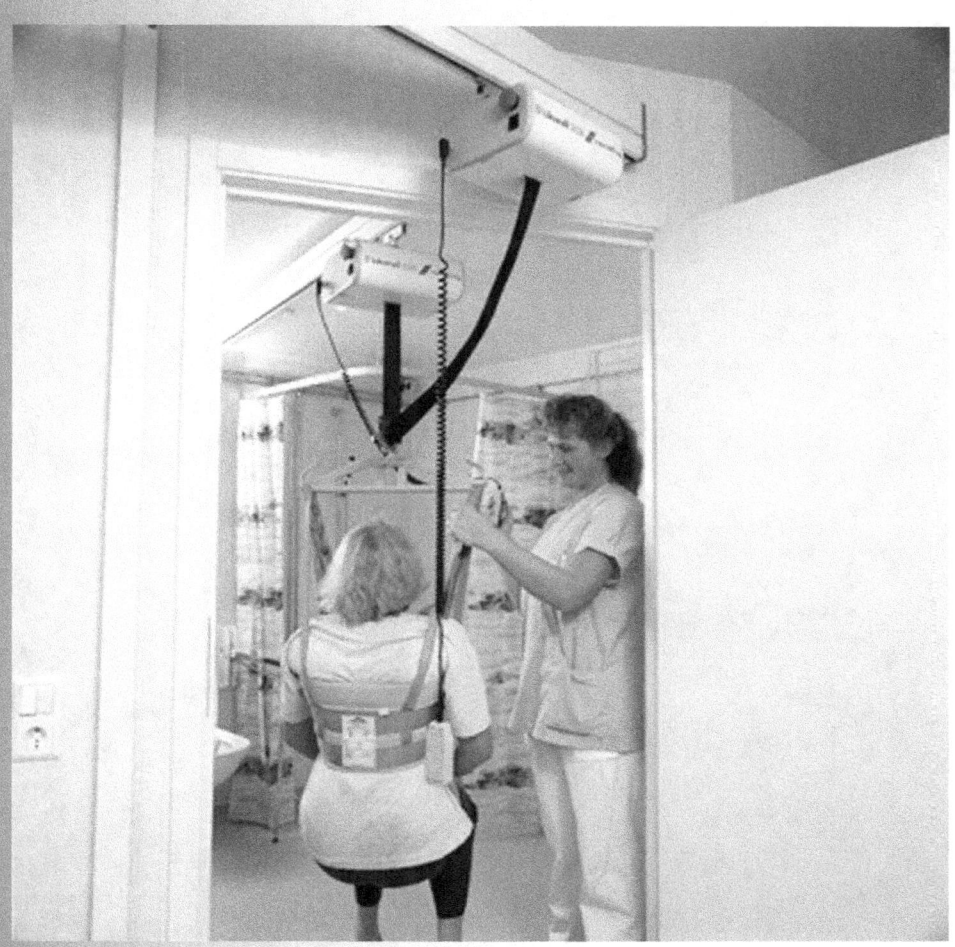

Safe Patient Handling Training for Schools of Nursing

Curricular Materials

Thomas R. Waters, Ph.D., NIOSH
Audrey Nelson, Ph.D., VHA
Nancy Hughes, Ph.D., ANA
Nancy Menzel, Ph.D., University of Las Vegas, NV

Curriculum developed in partnership with the National Institute for Occupational Safety and Health (NIOSH), the Veterans Health Administration (VHA), and the American Nurses Association (ANA)N

November 2009

Disclaimer

This document is in the public domain and may be freely copied or reprinted.

Mention of any company or product does not constitute endorsement by the National Institute for Occupational Safety and Health (NIOSH). In addition, citations to Web sites external to NIOSH do not constitute NIOSH endorsement of the sponsoring organizations or their programs or products. Furthermore, NIOSH is not responsible for the content of these Web sites. All web addresses referenced in this document were accessible as of the publication date.

Ordering Information

To receive documents or other information about occupational safety and health topics, contact NIOSH at

1-800-CDC-INFO (1-800-232-4636)
TTY: 1-888-232-6348
E-mail: cdcinfo@cdc.gov
or visit the NIOSH Web site at
www.cdc.gov/niosh

For a monthly update on news at NIOSH, subscribe to NIOSH eNews by visiting
www.cdc.gov/niosh/eNews

DHHS (NIOSH) Publication No. 2009-127
November 2009

SAFER • HEALTHIER • PEOPLE™

Table of Contents

Background — Pg 6

Tool Kit for Safe Patient Handling and Movement Training Program — Pg 8

Implementing the Curriculum Module for the First Time — Pg 9

Using the Curriculum — Pg 9

Required Student Reading — Pg 10

Required Student Viewing and Listening (narrated slide show) — Pg 10

Required Student Laboratory Activities — Pg 11

Optional Background Didactic Material for Faculty — Pg 11

Optional Laboratory Materials — Pg 14

Quiz for Didactic SPH — Pg 15

Appendices

Appendix A - VHA Safe Patient Handling and Movement Algorithms — Pg 16

Appendix B - Assessment Criteria and Care Plan for Safe Patient Handling and Movement — Pg 38

Background

In the field of nursing, work-related musculoskeletal disorders (MSDs), such as back and shoulder injuries, persist as the leading and most costly U.S. occupational health problem (Nelson et al., 2009). A large body of evidence indicates that a substantial number of work-related MSDs reported by nurses are due to the cumulative effect of repeated manual patient-handling activities and work done in extreme static awkward postures. In a list of at-risk occupations for musculoskeletal disorders in 2007, nursing aides, orderlies, and attendants ranked first in incidence rate with a case rate of 252 cases per 10,000 workers, a rate seven times the national MSD average for all occupations. Emergency medical personnel ranked second, followed by laborers and material movers, ticket agents and travel clerks, and light and heavy truck drivers among the top six at-risk occupations [Department of Labor, Bureau of Labor Statistics (BLS), 2009]. The nursing occupation also typically ranks in the top ten in yearly incidence rate of sprain and strain injuries.

In most industries MSD injury rates have declined in recent years, yet MSD rates for nurses in the healthcare industry have not declined during the same period. Healthcare units at high risk for back and other injuries to caregivers have certain characteristics:

- History of frequent injuries
- High proportion of dependent patients
- Lack of use of lifting equipment in good repair
- Low staffing levels

The high physical demands associated with handling and moving patients—who are getting heavier as obesity rates in the United States climb—are probably the largest contributing factor to high rates of MSDs among practicing nurses. The aging of the workforce likely contributes to the problem; the average age of a registered nurse in the United States is approximately 47 years. Also contributing to the negative health consequences of manual handling is the shortage of nurses—Peter Buerhaus, a researcher at Vanderbilt University Medical Center, has estimated that there will be a shortage of 285,000 nurses by the year 2020 and 500,000 by the year 2050 in the US—likely resulting in longer work hours and more demanding schedules for practicing nurses.

More than 30 years of evidence has demonstrated that manual patient handling and relying on body mechanics is unsafe (Nelson et al., 2009). Furthermore, this evidence indicates that adoption of safe patient handling (SPH) techniques, where nurses use assistive equipment during transfers, is effective in reducing the incidence of MSDs related to the handling of patients (Nelson et al., 2009).

Educators at schools of nursing, however, continue to teach outdated techniques for patient handling. These approaches rely on "proper" body mechanics—when there really is no safe method to manually lift another adult human being. The two-person lift and the hook-and-toss methods persist as primary approaches taught to student nurses for lifting and moving patients (Nelson et al., 2009). Experts and advocates of safety in handling patients consider these techniques unsafe (Nelson et al., 2009).

Likely reasons that SPH techniques have not been widely accepted nor incorporated into fundamental nursing education include the fact that (1) the knowledge base for applying ergonomics to handling patients has only recently evolved and (2) available evidence-based teaching materials and resources targeted toward faculty are lacking. This SPH curricular material, developed by cooperative effort among the National Institute for Occupational Safety and Health (NIOSH), the Veterans Health Administration (VHA), and the American Nurses Association (ANA), will help instructors design training programs that encourage the use of safe approaches to handling patients and contribute to the prevention of MSDs.

The entire program is designed to be implemented as a 1- or 2-day training module in schools of nursing. Each element of the program is designed to support and reinforce the other elements. The model for the SPH training program is focused on student learning, faculty development in effective teaching and student assessment, and it is based on three primary concepts:

1. Opinion leader(s) (Dean or Instructors) incorporates the SPH training program into the fundamentals nursing course.
2. Instructor or school negotiates with practice settings to support safer patient-handling practices through purchasing equipment and instituting low-lift policies.
3. Nursing schools permanently adopt the SPH module as part of the curriculum.

The effectiveness of the SPH training program was evaluated in a field study by partner educators at 29 schools of nursing. Nelson et al. [2007] describe the findings from the evaluation study. (An overview of the process used to develop the curriculum is described in a paper by Menzel et al. [2007].)

Equipment vendors participated in the evaluation project by providing state-of-the-art equipment to use in the clinical component of the training curriculum. The vendors also often provided a trained staff person to help the school during the clinical training phase of the training implementation. The evaluation involved a pre- and post-design with a group of schools of nursing. Nurse educators at 26 nursing schools received curricular materials and training; nursing students received the evidence-based curriculum module. There were three control sites. Questionnaires were used to collect data on knowledge, attitudes, and beliefs about safe handling of patients for both nurse educators and students, before and after training. The authors found that knowledge about SPH concepts improved considerably among both nurse educators and students at intervention schools. Also, both groups at the intervention schools indicate they are more likely to use mechanical lifting devices when they are available.

With a new curriculum that addresses the prevention of biomechanical hazards of patient handling, nurse educators can impart this knowledge directly to students who will soon be entering the labor force as professional nurses. As such, nurse educators are pivotal in changing perceptions throughout the nursing and healthcare industry in order to place a higher value on prevention of MSDs through safe patient-handling methods. From pools of students come individuals who eventually take on roles and positions of leadership as directors of nursing, supervisors of nursing departments, unit charge nurses, and heads of committees (e.g., safety and health committee)—all with some degree of decision-making authority. Further, nursing students who become front-line workers may serve as important advocates among facility staffs to adopt safe patient-handling methods, equipment, and policies. The potential impact of this influence throughout the range of healthcare settings—from acute care hospitals to nursing homes—can be extensively diffuse.

To date, approximately 90 nurse educators have been trained to teach the new training materials. More than 1,500 students have been through the training program. Based on requests for the SPH training materials from numerous schools of nursing, within the next 5 years an estimated 60,000 additional students could be trained using the new materials. If the training materials are adapted for use by healthcare groups for retraining current nurses and other caregivers, many healthcare workers who currently perform patient-handling tasks may be positively affected by this training program. In the future, it is likely that healthcare facilities will adopt SPH programs. Having a highly trained workforce will increase the effectiveness of these programs.

The training program has four main objectives:

1. Provide evidence-based training on SPH to instructors at schools of nursing so that they can teach SPH methods to students.
2. Ensure that the training is sound and that the curriculum is effective in improving the knowledge, attitudes, and beliefs of the students.
3. Provide a full range of educational tools nursing educators can use to increase effectiveness of the training program (see Tool Kit for Safe Patient Handling and Movement Training Program).
4. Encourage all nursing educators at schools of nursing to use the evidence-based, safe-patient-handling curriculum module and recommended laboratory activities for nurse training.

The ultimate goal is to move students beyond simply knowing content to applying what they've learned in a clinical setting.

Safe Patient Handling and Movement Concepts

In order to clarify the difference between the terms "biomechanics" and "body mechanics," we have provided the following definitions. Biomechanics is the study of the mechanics of muscular activity, and how muscular activity leads to internal loading of body tissues, such as the ligaments, joints, and other soft tissues. Biomechanics is useful in determining whether a specific manual patient handling task will create unacceptably high forces inside the body and whether a manual lift is "safe" or not. Body mechanics, on the other hand, is a belief that reliance on "correct" body positions or "body movements" will somehow provide protection from the force associated with lifting and moving patients. Body mechanics is also used to assess the alignment of patients when they are standing, sitting, or lying down. Body mechanics alone, however, is not sufficient to protect the nurse from the heavy weight, awkward postures, and repetition involved in manual handling. Safe manual handling techniques must be used in combination with equipment and technology for safe patient handling and movement. There are four primary principles of manual patient handling that should be used *in conjunction with* SPH techniques when handling and moving patients. The four principles include:

The four principles include:

1. Maintain a wide, stable base with your feet.

2. Put the bed at the correct height (waist level when providing care; hip level when moving a patient.)

3. Try to keep the work directly in front of you to avoid rotating the spine.

4. Keep the patient as close to your body as possible to minimize reaching.

Preparing for the Patient Handling Activity

1. Take responsibility for knowing how equipment works and whether it's available.

2. Assess the client and the environment using the Assessment Criteria and Care Plan for Safe Patient Handling and Movement.

3. Select the appropriate algorithm.

4. Gather the appropriate equipment and other staff members, if needed.

5. Organize the physical environment and the equipment to ensure safe completion of the task. This includes locking the wheels of the bed or chair, putting the bed or stretcher at the correct height, removing clutter, and making sure any mobile equipment is charged.

6. Make sure other team members, if any, know their roles; rehearse if necessary.

7. Position yourself using the principles of body mechanics (above).

8. Coach the patient.

9. Tell patients what actions you the student plan and expect from them. Show them what to do, and then help them move through the activity.

Tool Kit for Safe Patient Handling and Movement Training Program

This tool kit contains didactic content and clinical laboratory content required to incorporate evidence-based patient handling into a nursing-school curriculum.

The curriculum consists of four main elements:

1. A narrated, approximately 2-hour slide presentation

2. A series of algorithms (i.e., decision tools that help nurses assess patient needs to decide which equipment is appropriate for a specific patient-handling activity)

3. Didactic materials

4. Laboratory activities

For more extensive information about SPH, consult the 2006 publication *Handle with Care: A Practice Guide for Safe Patient Handling and Movement,* A. Nelson editor, or, the 2009 publication *An Illustrated Guide to Safe Patient Handling and Movement,* A Nelson, K. Motacki, and N. Menzel.

After taking training based on this curriculum, students should be able to:

- define healthcare ergonomics;

- recognize high-risk, patient-care activities;

- identify risks in patient-care environments;

- state why mechanical aids are needed when moving and handling patients;

- use algorithms to identify safe patient-handling and movement strategies;

- assess patients to select the right combination of equipment and personnel needed to handle or move them safely;

- apply positioning and mobility techniques that are

safe for patient and caregiver.

At the end of this material is a quiz students can take that will help them evaluate their knowledge of new patient-handling concepts.

Implementing the Curriculum Module for the First Time

1. Be proactive.

    - Organize local, focused workshops, for example, on sling selection, that are accessible to faculty and staff nurses.

    - Send flyers to local hospitals to announce your (instructor) involvement in SPH.

    - Conduct outreach to local facilities, such as clinical educators in local hospitals and nursing homes, local nurse executives, and hospital CEOs.

    - Hold an open house/equipment fair for local hospitals.

2. Ensure that all faculty members receive training and provide continued emphasis on lack of effectiveness of reliance on body mechanics alone.

3. Invest in developing collaborative efforts with vendors; equipment should be housed on site.

4. Attend the annual SPH conference sponsored by the Veterans Integrated Service Network 8 (VISN 8) Patient Safety Center of Inquiry and the University of South Florida.

5. Encourage students to talk about SPH to hospital staff.

6. Emphasize to students the evidence underlying the program.

7. Use SPH program to illustrate evidence-based practice in other courses.

8. Obtain equipment early to ensure availability at start of session and during preparation, insure that equipment is maintained properly, and change equipment as necessary.

Informing Others about the benefits of the SPH Curriculum

- Develop a speaker's bureau with members who can speak about the program to others interested in implementation.

- Develop applications for specialty clinical areas such as critical care, emergency department, and obstetrics.

- Conduct outreach to state nursing organizations, state workforce and labor offices, and other schools of nursing.

- Develop a sample policy statement for student handbooks about SPH for adoption by colleges of nursing.

- Get involved in changing nursing textbooks and in writing National Council Licensure Examination-Registered Nurse (NCLEX) questions on SPH.

Using the Curriculum

The first step in implementation of the curriculum is to introduce the students to the problem and solutions using a series of didactic activities. The didactic activities are followed by a hands-on laboratory session. The didactic and laboratory content is described below.

Didactic

1. Assign required readings to students. Provide reserve copies of these copyrighted print articles or links to them via the school's library.
2. Instruct the students to print and have at hand the Assessment Criteria and Care Plan for Safe Patient Handling and Movement and the algorithms; they should have those available when they watch the slide show (see Required Student Viewing and Listening below).
3. Assign viewing of the SPH Curriculum Module slide show either on the student's own time or in class.
4. Consider administering a sample quiz similar to that included with the optional materials.

Laboratory

Once the students have completed the didactic portion of the module, they should have an opportunity to practice SPH in a laboratory setting. This will require at least the following types of equipment: ceiling lift, mobile lift, sit-to-stand lift, and a friction-reducing device. Your school might purchase or seek a donation of these devices or borrow them from a practice partner or a vendor.

Students should have access to the reading resources listed below...

Required Student Reading

Myths and Facts About Back Injuries in Nursing by A. Nelson et al. in *The American Journal of Nursing*, Vol 103(2), pages 32–41.

Safe Patient Handling and Movement by A. Nelson et al. in the *The American Journal of Nursing*, Vol 103(2), pages 32–43.

The Illustrated Guide to Safe Patient Handling and Movement by A. Nelson et al., K. Motaki, and N. Menzel, 2009, Springer Publishing.

When is it Safe to Manually Lift a Patient? By T. Waters in the *American Journal of Nursing*, Vol 107(8), pages 53-59.

(Since some of these articles are copyrighted, each school should make these articles available for students, either on reserve or electronically.)

ANA Position Statement on safe patient handling to prevent work-related musculoskeletal disorders. Accessible from the following hyperlink: [http://nursingworld.org/MainMenuCategories/OccupationalandEnvironmental/occupationalhealth/handlewithcare.aspx]

Assessment Criteria and Care Plan for Safe Patient Handling and Movement and the Algorithms. Available in Appendix B of this document and downloadable from the following internet hyperlink: www.visn8.med.va.gov/visn8/patientsafetycenter/safePtHandling/default.asp. (At the web site, select Assessment Form and Algorithms under the heading Algorithms for Safe Patient Handling and Movement.)

Required Student Viewing and Listening

One of the main components of the training program is a narrated Adobe Flash multimedia slideshow titled, Safe Patient Handling and Movement Principles.

You can also download the SPH training presentation from the NIOSH web site.

This presentation requires that students have in front of them copies of the Assessment Criteria and Care Plan for Safe Patient Handling and Movement and the algorithms which are contained in this booklet.

Instructions for downloading the materials are available at the NIOSH web site. www.cdc.gov/NIOSH.

Recommended Student Laboratory Activities

1. Students will be trained in the operation of SPH aids.
2. Students will practice assessing "patients" (mannequins or other students) for their handling and movement needs using the Assessment Criteria and Care Plan for Safe Patient Handling and Movement. Faculty will set up stations representing a variety of tasks that should include, at minimum, the following:
   - Lift fully dependent patient out of bed and into chair using powered full-body sling lift (overhead or mobile)
   - Assist cooperative patient with minimal lower-body strength from sitting position to standing position using powered sit-to-stand lift
   - Transfer fully dependent patient to stretcher from bed using lateral transfer device, such as a friction-reducing device
   - Reposition a patient in bed (side-to-side, up-in-bed) using a friction-reducing device

Optional Background Didactic Material for Faculty

Principles of Safe Patient Handling and Movement

Patient-handling and movement activities are a necessary part of basic nursing care. Nurses of all ages and experience levels become injured on the job while performing tasks such as getting patients out of bed, transferring them to stretchers, or pulling them up in bed. The way to decrease the occurrence of the MSDs does not lie in improving manual lifting techniques or hiring stronger nurses but in assessing what is needed to accomplish the task safely and then carrying out the task appropriately. Many tasks require the use of mechanical equipment or other SPH aids. This is called ergonomic intervention or modifying the job to protect the worker.

Symptoms of MSDs include pain that varies according to stage:
- Early stage—Pain may disappear after a rest away from work.
- Intermediate stage—Body part aches and feels weak soon after starting work and lasts until well after finishing work.
- Advanced stage—Body part aches and feels weak even at rest, sleep may be affected, light tasks are difficult on days off.

Other symptoms include tingling or numbness, fatigue, and weakness. Signs include redness and swelling or loss of full or normal joint movement.

Individuals should not ignore signs and symptoms of MSDs. For optimum treatment effectiveness and to reduce the risk of severe health outcomes, people should report signs and symptoms as soon as possible. If employees experience signs or symptoms of musculoskeletal disorders, they should report them to the safety and health department as early as possible and seek treatment.

The basic principles of ergonomics, or fitting the job to the worker, seem to offer the best hope for improving the problems associated with muscle and joint disorders caused by nursing tasks. The ergonomic approach is a step toward the goal of decreasing the number and severity of job-related injuries in nursing practice by increasing safety and decreasing fatigue. Patient-care ergonomics will improve productivity and help caregivers feel less tired at the end of the workday.

Ergonomic approaches are used to
1. design jobs and job tasks to fit each person rather than expecting each person to adapt to poor work designs;
2. achieve a proper match between the worker and his or her job by understanding and incorporating the limits of the individual
3. evaluate the work environment, taking into account that when job demands exceed the physical ability of workers, problems likely exist.

Risk Factors that May Lead to Musculoskeletal Disorders

The key to nurses working safely is careful analysis of the risk factors involved in providing patient care. Ergonomic assessments focus on two of these key factors: characteristics of the work environment and characteristics of job tasks.

Characteristics of the Work Environment

There is greater risk of injury for nurses in nursing homes, geriatric units, and spinal cord injury units than in general hospital units because more patient handling occurs in these work environments. Characteristics of the work environment can include the nursing practice setting, patient assignments, scheduling, space,

equipment, and staffing. The environment offers many factors that increase the risk of injury for nurses:
- Slippery or wet surfaces
- Physical obstructions (e.g., cabinets, toilets)
- Obstructions on floor surfaces
- Uneven floor surfaces
- Uneven work surfaces (e.g., different heights between caregivers arms and bed, different heights between wheelchair and toilets)
- Too small or too difficult-to-access spaces
- Too small a width to the entrance way
- Poor arrangement of furnishings
- Poor bathing area design
- Poor design of chairs

Much of the problem regarding back injuries to workers in nursing facilities has been blamed on lack of appropriate equipment in working condition and on understaffing. Some lifts and patient care tasks do require multiple staff members to accomplish safely, and, because low staffing makes this teamwork difficult, nurses often attempt these tasks alone.

Characteristics of Job Tasks

The risk for lifting injuries increases for nurses who hold patients away from the body while lifting and when they bend and twist during the lifting process. This awkward angle and position frequently occurs during bathing and feeding, but the greatest risk is associated with one-person transferring techniques. Sudden high effort from unexpected events, such as preventing a patient from falling, is also associated with high risk for injury.

A caregiver commonly performs on-the-job activities that put him or her at risk of an MSD:
- Reaching and lifting loads far from the body
- Lifting heavy loads (greater than 40–50 pounds under ideal conditions)
- Twisting while lifting
- Reaching low or high to begin a lift
- Moving a load a great distance
- Frequent lifting (more than 12 lifts a shift)
- Unassisted lifting
- Awkward posture of caregiver

Job tasks may have characteristics that are beyond the caregiver's control that put him or her at risk of an MSD.

- Unexpected changes during the lift (e.g., combative patient, falling patient)
- Excessive pushing or pulling forces required to accomplish task
- Lack of ability to grasp the patient securely (no handles)
- Totally dependent, unpredictable, or combative patient
- Patient's inability to understand
- Patient's special medical conditions such as burns or stroke

Some nursing tasks put the caregiver at risk of injury if done without assistance:
- Transferring patient from bathtub to wheelchair, wheelchair to shower/commode chair, wheelchair to bed, bed to stretcher, and vice versa
- Lifting a patient from the floor
- Weighing a patient
- Bathing a patient in bed, in a shower chair, or on a shower trolley or stretcher
- Undressing/dressing a patient, including applying antiembolism stockings
- Repositioning patient in bed from side to side or to the head of the bed
- Repositioning patient in geriatric chair or wheelchair
- Making an occupied bed
- Feeding a bed-ridden patient
- Changing absorbent pad when bed is occupied

Summary

In the past, nursing schools taught that body mechanics (i.e., specialized positioning of the caregiver's body) could protect against injury during manual handling procedures. However, research has shown that body mechanics alone is not protective:
- Manual lifting techniques are not based on studies of women.
- Manual lifting techniques focus only on the lower back.
- Manual lifting techniques are based on loads that weigh far less than typical patients.
- Manual lifting techniques are based on the assumption that the load is stable and held close to the body.

In addition, many of the characteristics of moving patients make body mechanics unsuitable for manual

handling:
- The load is often unstable.
- Patients do not have handles.
- A patient's weight is distributed unevenly.
- A patient may be combative.

From *The Nurse's Load* (Lancet editorial, 1965, p.422). "The adult human form is an awkward burden to lift or carry. Weighing up to 100 kilograms or more, it has no handles, it is not rigid, and it is liable to severe damage if mishandled or dropped. In bed a patient is placed inconveniently for lifting, and the placing of a load in such a situation would be tolerated by few industrial workers. . . . Since much of the nurse's day is spent in lifting patients, it is no small wonder that orthopedic wards often contain nurses with strained backs as patients."

Safe Patient-handling and Movement Aids

The key to effective back injury prevention programs is the use of SPH approaches that analyze job tasks and identify risk factors with the purpose of changing unacceptable job demands. SPH aids include the following:
- Powered, mechanical full-body lifts (either mobile or ceiling-mounted)
- Powered, mobile sit-to-stand lifts
- Friction-reducing devices
- Transfer belts

These aids work by bearing most of the load, reducing the load by lowering the friction between skin against cloth.

A Technology Resource Guide listing many types of patient handling equipment is available at the VA VISN8 Safe Patient Handling web site mentioned previously and listed below:
http://www.visn8.med.va.gov/patientsafetycenter/safePtHandling/default.asp.

Besides ergonomic interventions, other changes are needed to maximize caregiver safety:
- A clutter-free bedside environment to allow for free movement of equipment and personnel
- Coworker attitudes supportive of ergonomic interventions
- Adequate supply of modern SPH equipment conveniently located and in good working order
- SPH equipment that does not markedly slow the care process
- Administrative support through the use of a nonpunitive no-lift or low-lift policy
- Safety concern reporting system with rapid follow up
- Supervisor encouragement of early reporting for MSDs

Ergonomic interventions alleviate musculoskeletal discomforts for caregivers, yet patients also realize physical, psychological, and quality-of-life benefits:
- Increased patient comfort, security, and dignity during lifts and transfers
- Fewer patient falls, skin tears, or abrasions during transfers
- Promotion of patient movement and independence because the patient regains their strength sooner
- Enhanced toileting outcomes and increase in continence
- Improved overall quality of life

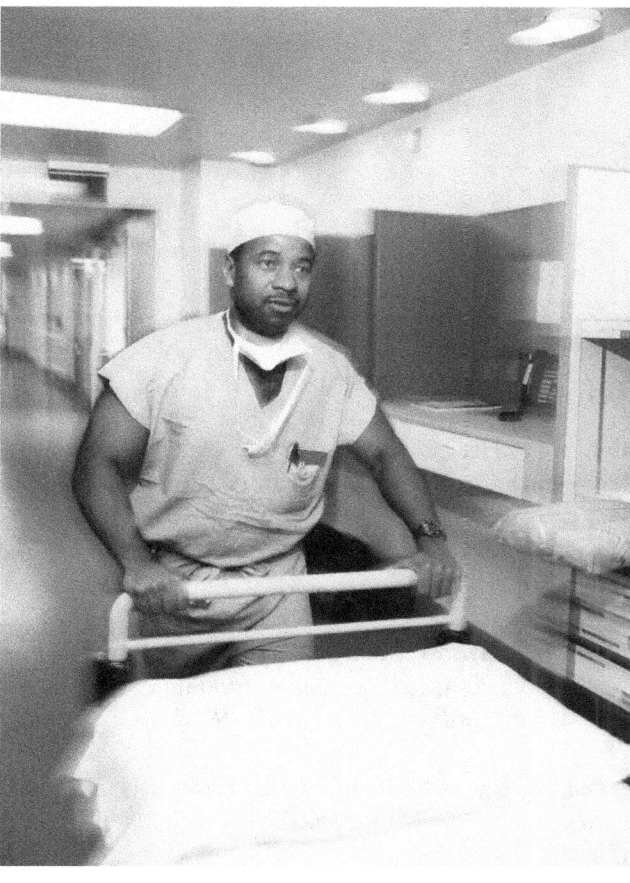

Optional Laboratory Materials

Skill	Proficient	Not Proficient
Assesses patient's handling and movement needs, demonstrated by selecting the correct algorithm		
Selects correct SPH aid or specifies number of other staff members needed to complete activity		
Describes or demonstrates proper operation of SPH aid prior to use with patient		
Positions patient correctly for use of SPH aid		
Arranges environment to allow use of SPH aid		
Applies SPH aid to patient correctly		
Completes SPH activity safely for self and patient		

References

Menzel N, Hughes N, Waters T, Shores L, Nelson A [2007]. Preventing musculoskeletal disorders in nurses: developing a safe patient handling curriculum module for schools of nursing. *Nurse Educ* 32(3):130–135.

Nelson A, Motaki K, and Menzel N. [2009] *The Illustrated Guide to Safe Patient Handling and Movement,* Springer Publishing.

Nelson A, Waters T, Menzel N, Hughes N, Hagan P, Powell-Cope G, Sedlak C, Thompson V [2007]. Effectiveness of an evidence-based curriculum module in nursing schools targeting safe patient handling and movement. *Int J Nurs Educ Scholarsh* 4(1): Article 26.

Bureau of Labor Statistics (BLS) [2009], US Department of Labor, Musculoskeletal disorders and days away from work in 2007. Accessed May 7, 2009 at http://data.bls.gov/cgi-bin/print.pl/opub/ted/2008/dec/wk1/art02.htm.

Quiz for Didactic SPH

Select the one best answer to these multiple choice questions.

1. Ergonomics means
    a. Making changes to the job to fit the worker*
    b. Making changes to the worker to fit the job
    c. Making workers work harder at their job
    d. Selecting stronger workers for the job

2. The goal of patient care ergonomics is to
    a. Slow down your work
    b. Help you feel and work better*
    c. Increase your work load
    d. Make patients recover faster

3. Which of the following patient care tasks involve heavy lifting?
    a. Charting
    b. Talking with the patient
    c. Transferring an immobile patient*
    d. Giving medications

4. Which of the following is a work environment factor that can reduce safety for both patient and caregiver?
    a. Caregiver educational level
    b. Uneven work surfaces*
    c. Patient BMI above 40
    d. Safe lifting policy

5. While bending forward, you spend 30 minutes feeding a patient on bed rest. What is (are) the musculoskeletal risk factor(s) in this situation?
    a. Pushing/pulling
    b. Awkward posture
    c. Long duration
    d. Heavy lifting
    e. b and c*
    f. All of the above

6. The purpose of assessing tasks and surroundings for risk factors is to
    a. Take steps to protect yourself*
    b. Slow down your work pace
    c. Delay care to the patient
    d. Distribute the workload to staff

7. If you had to transfer a totally dependent patient from a nonadjustable stretcher to a nonadjustable bed of different heights, what is the best step you could take to reduce the musculoskeletal risk factors?
    a. Use a friction-reducing device when transferring*
    b. Use a wide base of support when transferring
    c. Coach the patient to make the transfer unaided
    d. Use a draw sheet to transfer the patient

8. A staff nurse asks you to help her perform a lift you feel is unsafe. What would be your best response?
    a. "I'm busy caring for another patient, but I will help find someone to assist."
    b. "What does the safe lifting algorithm say we need to move the patient?"
    c. "Let me check with my instructor if I am allowed to help lift this patient."*
    d. "Tell me how you would like me to assist you with moving the patient."

9. Why are mechanical aides needed for patient handling?
    a. Nurses do not have sufficient training using proper body mechanics.
    b. Manual lifting techniques are not sufficient to protect nurses from injury.*
    c. Body mechanics algorithms are too complicated and difficult to understand.
    d. Nursing staff levels have declined in most institutions in recent years.

10. Use of a gait belt reduces what risk factor from moving patients that isn't present when moving boxes?
    a. Weight
    b. Dependence
    c. Cooperation
    d. No handles*

* Indicates correct answer.

Appendices

Appendix A- VHA Safe Patient Handling and Movement Algorithms

The Assessment Criteria and Care Plan for Safe Patient Handling and Movement, the algorithms, and the clinical tools that follow are taken directly from the Web site of the Department of Veterans Affairs: www.visn8.med.va.gov/visn8/patientsafetycenter/safePtHandling/default.asp

Algorithm 1: Transfer to and From: Bed to Chair, Chair to Toilet, Chair to Chair, or Car to Chair
Last rev. 10/012008

Start Here

Can patient bear weight?
- Fully → Caregiver assitance not needed; Stand by for safety as needed.
- Partially → Is the patient cooperative?
  - Yes → Stand-and-pivot technique using a gait/transfer belt (1 caregiver) or powered stand-assist lift (1 caregiver).
  - No → Use full-body sling lift and 2 caregivers.
- No → Is the patient cooperative?
  - Yes → Does the patient have upper-extremity strength?
    - Yes → Seated transfer aid; may use gait/transfer belt until the patient is proficient in completing transfer independently.
    - No → Use full-body sling lift and 2 caregivers.

- For seated transfer aid, must have chair with arms that recess or are removable.
- For full body sling lift, select a lift that was specifically designed to access a patient from the car (if the car is the starting or ending destination).
- If patient has partial weight-bearing capacity, transfer toward stronger side.
- Toileting slings are available for toileting.
- Mesh slings are available for bathing.
- During any patient-transferring task, if any caregiver is required to lift more than 35 lbs of a patient's weight, then the patient should be considered to be fully dependent and assistive devices should be used for the transfer. (Waters, T. [2007]. When is it safe to manually lift a patient? *American Journal of Nursing, 107*[8], 53-59.)

Algorithm 2: Lateral Transfer To and From: Bed to Stretcher, Trolley
Last rev. 01/13/2009

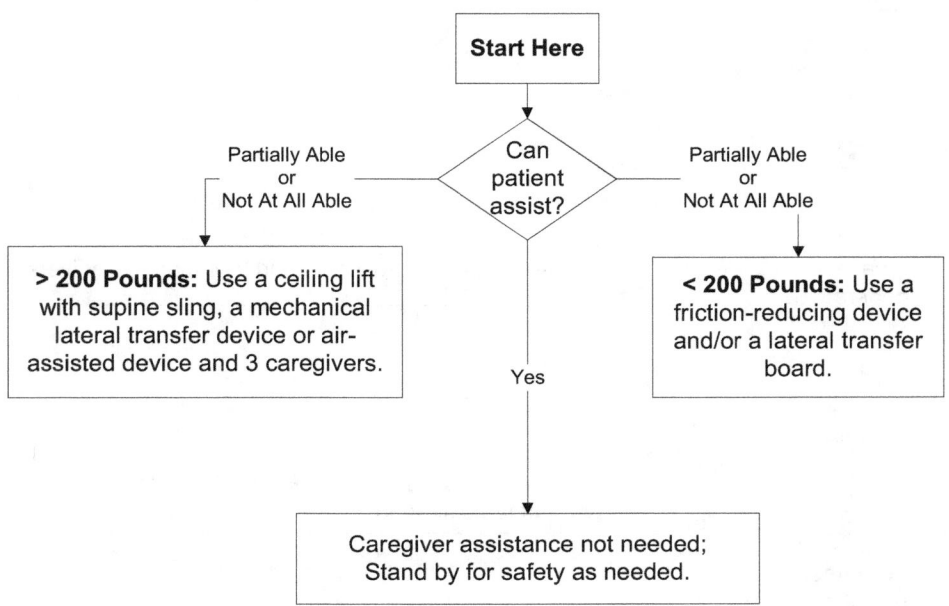

- Destination surface should be 1/2" lower for all lateral patient moves.
- For patients with Stage III or IV pressure ulcers, care must be taken to avoid shearing force.
- During any patient transferring task, if any caregiver is required to lift more than 35 lbs of a patient's weight, then then patient should be considered to be fully dependent and assistive devices should be used for the transfer. (Waters, T. [2007]. When is it safe to manually lift a patient? *American Journal of Nursing, 107*[8], 53-59.)

Algorithm 3: Transfer To and From: Chair to Stretcher or Chair to Exam Table
Last rev. 10/01/08

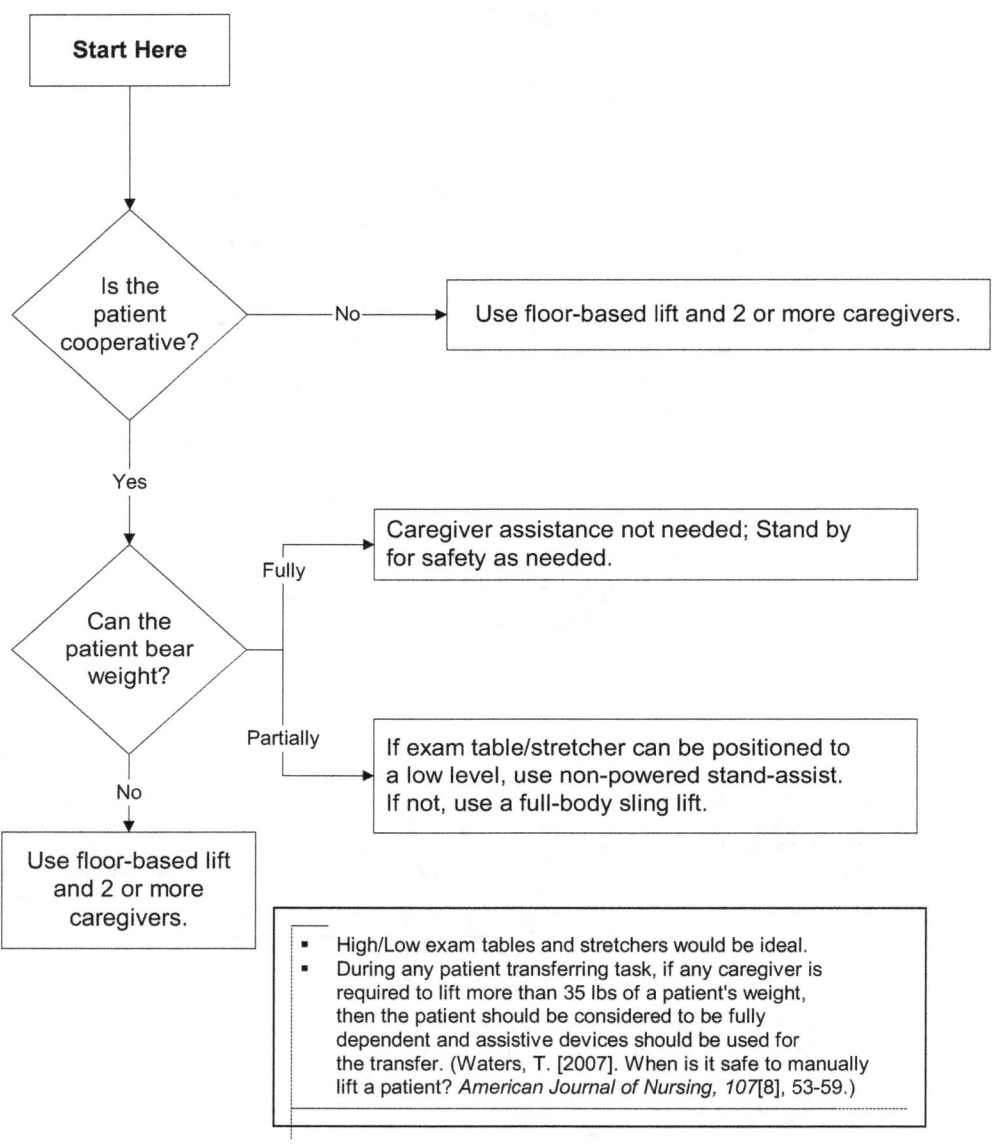

Algorithm 4: Reposition in Bed: Side-to-Side, Up in Bed
Last rev. 10/01/08

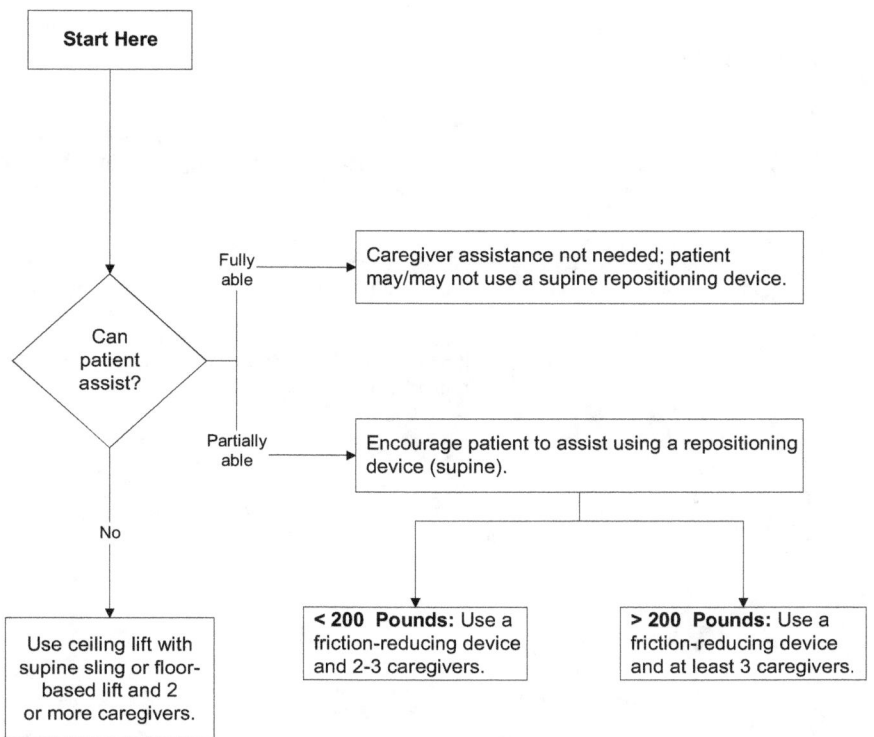

- This is not a one person task: DO NOT PULL FROM HEAD OF BED.
- When pulling a patient up in bed, the bed should be flat or in a Trendelenburg position (when tolerated) to aid in gravity, with the side rail down.
- For patients with Stage III or IV pressure ulcers, care should be taken to avoid shearing force.
- The height of the bed should be appropriate for staff safety (at the elbows).
- If the patient can assist when repositioning "up in bed," ask the patient to flex the knees and push on the count of three.
- During any patient handling task, if the caregiver is required to lift more than 35 lbs of a patient's weight, then the patient should be considered to be fully dependent and assistive devices should be used.
  (Waters, T. [2007]. When is it safe to manually lift a patient? *American Journal of Nursing, 107*[8], 53-59.)

Algorithm 5: Reposition in Chair: Wheelchair and Dependency Chair
Last rev. 10/01/08

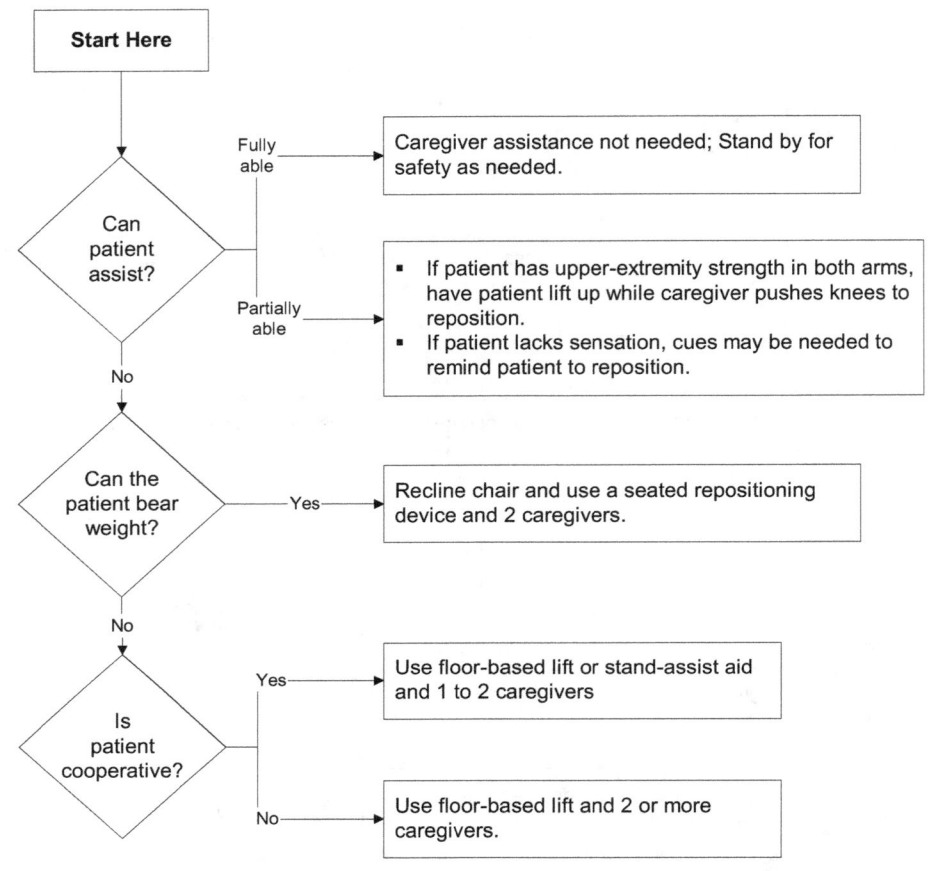

- Take full advantage of chair functions, e.g., chair that reclines, or use arm rest of chair to facilitate repositioning.
- Make sure the chair wheels are locked.
- During any patient transferring task, if any caregiver is required to lift more than 35 lbs of a patient's weight, then the patient should be considered to be fully dependent and assistive devices should be used. (Waters, T. [2007]. When is it safe to manually lift a patient? *American Journal of Nursing, 107*[8], 53-59.)

Algorithm 6: Transfer a Patient Up From the Floor
Last rev. 10/01/08

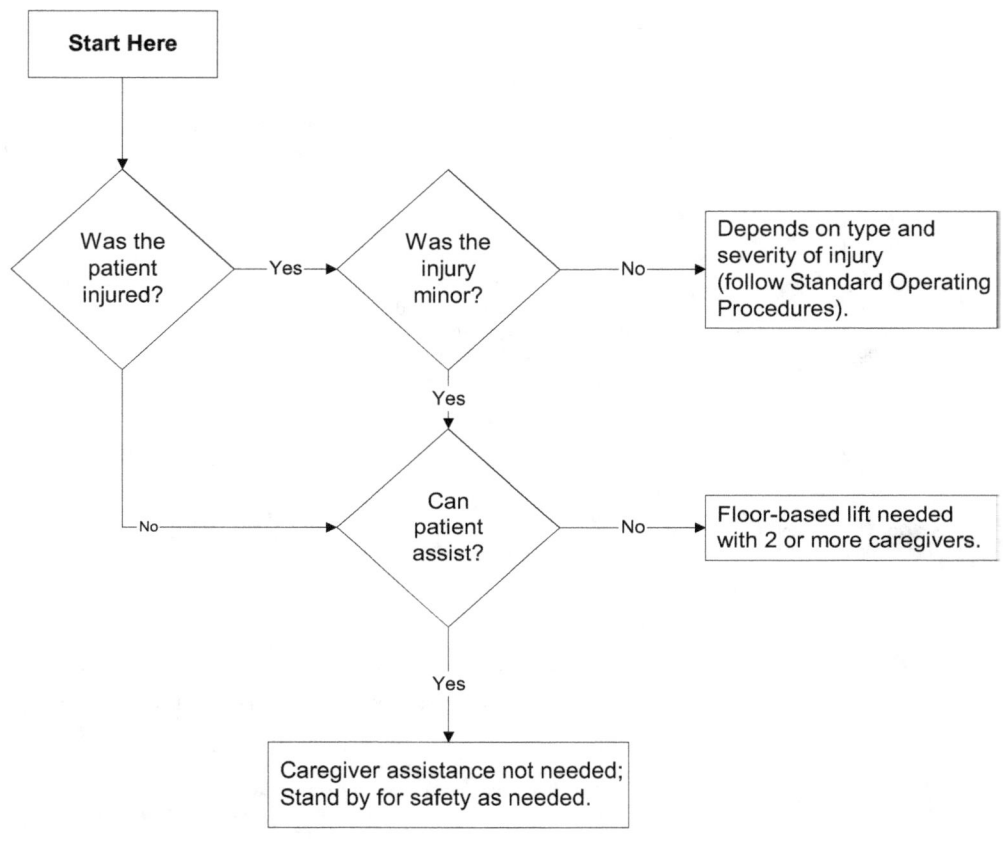

- Use floor-based lift that goes all the way down to the floor (most of the newer models are capable of this).
- During any patient transferring task, if any caregiver is required to lift more than 35 lbs of a patient's weight then the patient should be considered to be fully dependent and assistive devices should be used. (Waters, T. [2007]. When is it safe to manually lift a patient? *American Journal of Nursing, 107*[8], 53-59.)

Bariatric Algorithm 1: Bariatric Transfer To and From: Bed/Chair, Chair/Toilet, or Chair/Chair
rev. 10/01/08

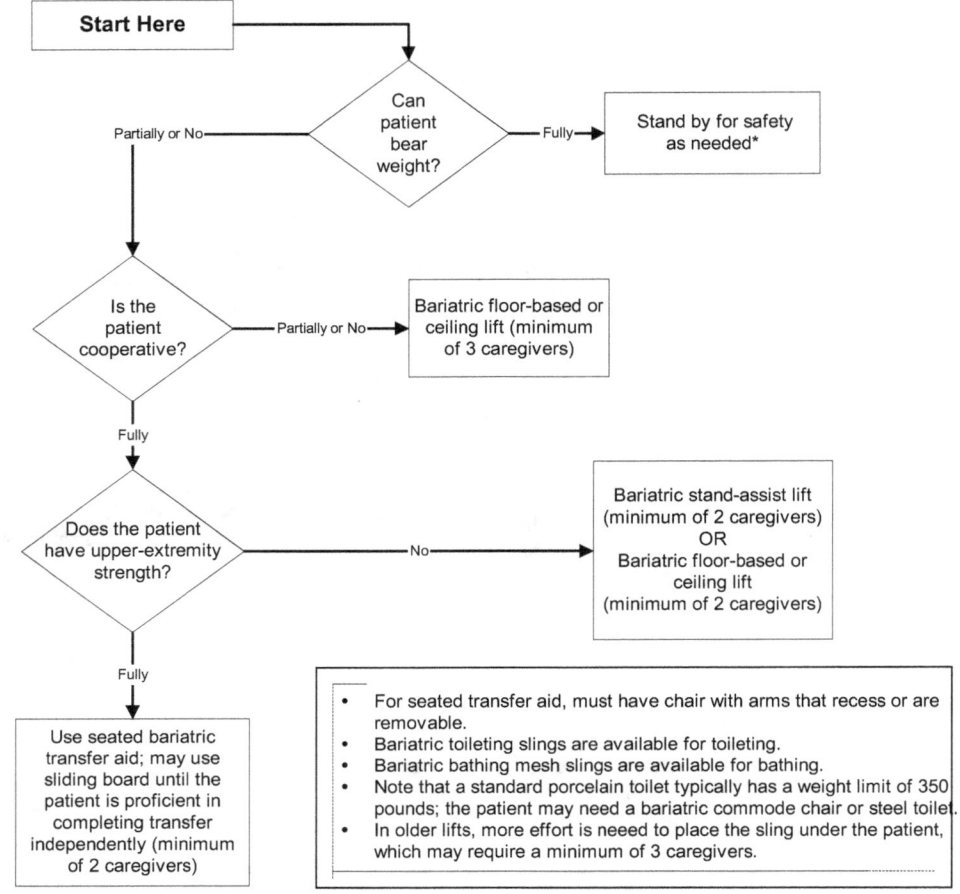

- "Stand by for safety." In most cases, if a bariatric patient is about to fall, there is very little that the caregiver can do to prevent the fall. The caregiver should be prepared to move any items out of the way that could cause injury, try to protect the patient's head from striking any objects or the floor and seek assistance as needed once the person has fallen.
- If patient has partial weight-bearing capability, transfer toward stronger side.
- Consider using an abdominal binder if the patient's abdomen impairs a patient-handling task.
- Assure equipment used meets weight requirements. Standard equipment is generally limited to 250-350 lbs. Facilities should apply a sticker to all bariatric equipment with "EC" (for expanded capacity) and a space for the manufacturer's rated weight capacity for that particular equipment model.
- Identify a leader when performing tasks with multiple caregivers. This will assure that the task is synchronized for increased safety of the health care provider and the patient.
- During any patient transferring task, if any caregiver is required to lift more than 35 lbs of a patient's weight, then the patient should be considered to be fully dependent and assistive devices should be used. (Waters, T. [2007]. When is it safe to manually lift a patient? American Journal of Nursing, 107[8], 53-59.)

Bariatric Algorithm 2: Bariatric Lateral Transfer To and From: Bed/Stretcher/Trolley
rev. 10/01/08

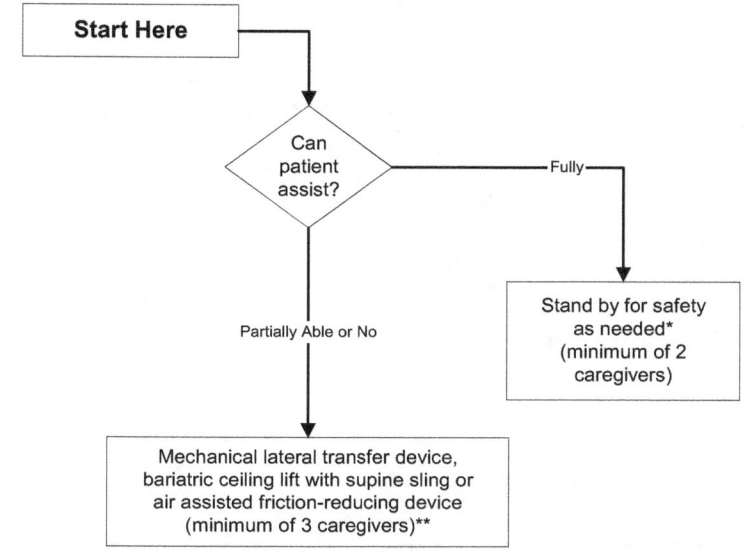

- The destination surface should be about 1/2" lower for all lateral patient moves.
- Avoid shearing force.
- Make sure bed is the right width, so excessive reaching by caregiver is not required.
- Lateral transfers should not be used with speciality beds that interfere with the transfer. In this case, use a bariatric ceiling lift with supine sling.
- Ensure bed or stretcher doesn't move with the weight of the patient transferring.
** Use a bariatric stretcher or trolley if patient exceeds weight capacity of traditional equipment.

- * "Stand by for safety." In most cases, if a bariatric patient is about to fall, there is very little that the caregiver can do to prevent the fall. The caregiver should be prepared to move any items out of the way that could cause injury, try to protect the patient's head from striking any objects or the floor and seek assistance as needed once the person has fallen.
- * Assure equipment used meets weight requirements. Standard equipment is generally limited to 250-350 lbs. Facilities should apply a sticker to all bariatric equipment with "EC" (for expanded capacity) and a space for the manufacturer's rated weight capacity for that particular equipment model.
- If patient has partial weight-bearing capability, transfer toward stronger side.
- Consider using an abdominal binder if the patient's abdomen impairs a patient-handling task.
- Identify a leader when performing tasks with multiple caregivers. This will assure that the task is synchronized for increased safety of the health care provider and the patient.
- During any patient transferring task, if any caregiver is required to lift more than 35 lbs of a patient's weight, then the patient should be considered to be fully dependent and assistive devices should be used. (Waters, T. [2007]. When is it safe to manually lift a patient? *American Journal of Nursing, 107*[8], 53-59.)

Bariatric Algorithm 3: Bariatric Reposition in Bed: Side-to-Side, Up in Bed
rev. 10/01/08

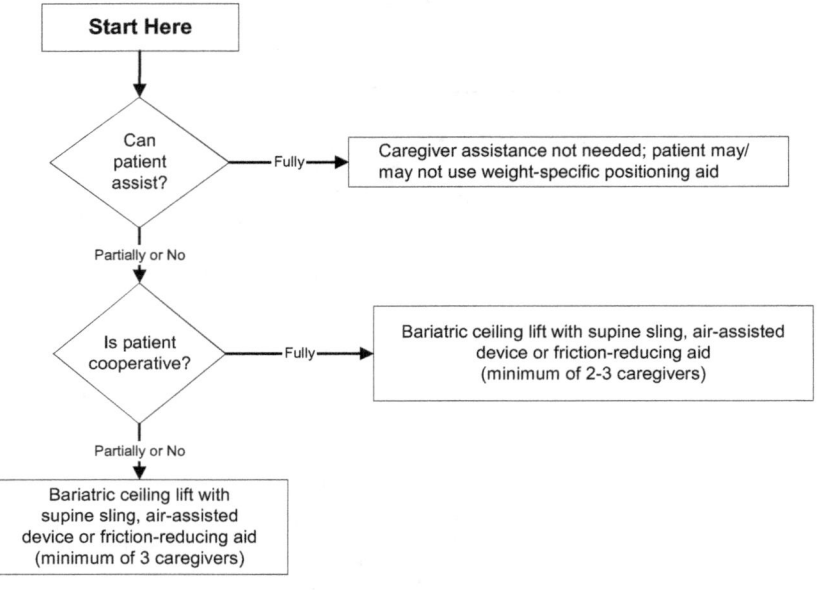

- When pulling a patient up in bed, place the bed flat or in a Trendelenburg position (if tolerated and not medically contraindicated) to aid in gravity; the side rail should be down.
- Avoid shearing force.
- Adjust the height of the bed to elbow height.
- Mobilize the patient as early as possible to avoid weakness resulting from bed rest. This will promote patient independence and reduce the number of high-risk tasks caregivers will provide.
- Consider leaving a friction-reducing device covered with drawsheet, under patient at all times to minimize risk to staff during transfers as long as it doesn't negate the pressure relief qualities of the mattress/overlay.
- Use a sealed, high-density, foam wedge to firmly reposition patient on side. Skid-resistant texture materials vary and come in set shapes and cut-your-own rolls. Examples include:
  - Dycem (TM)
  - Scoot-Guard (TM): antimicrobial; clean with soap and water, air dry.
  - Posey-Grip (TM): Posey-Grip does not hold when wet. Washable, reusable, air dry.

- If patient has partial weight-bearing capability, transfer toward stronger side.
- Consider using an abdominal binder if the patient's abdomen impairs a patient-handling task.
- Assure equipment used meets weight requirements. Standard equipment is generally limited to 250-350 lbs. Facilities should apply a sticker to all bariatric equipment with "EC" (for expanded capacity) and a space for the manufacturer's rated weight capacity for that particular equipment model.
- Identify a leader when performing tasks with multiple caregivers. This will assure that the task is synchronized for increased safety of the healthcare provider and the patient.
- During any patient transferring task, if any caregiver is required to lift more than 35 lbs of a patient's weight, then the patient should be considered to be fully dependent and assistive devices should be used. (Waters, T. [2007]. When is it safe to manually lift a patient? *American Journal of Nursing, 107*[8], 53-59.)

Bariatric Algorithm 4: Bariatric Reposition in Chair: Wheelchair, Chair, or Dependency Chair
rev. 10/01/08

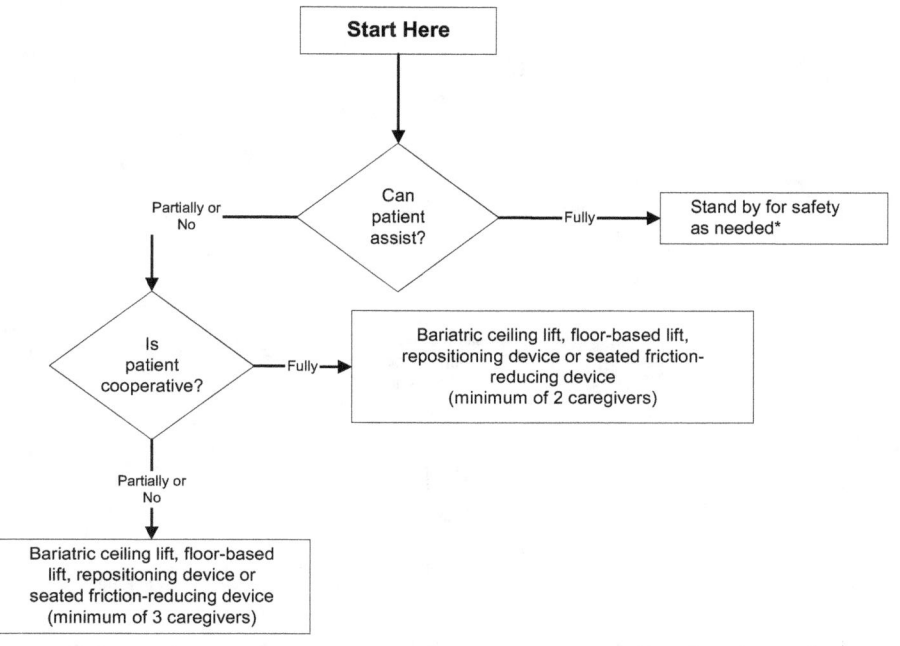

- Take full advantage of chair functions, e.g., chair that reclines, or use an arm rest of chair to facilitate repositioning.
- Make sure the chair wheels are locked.
- Consider leaving the sling under the patient at all times to minimize risk to staff during transfers after carefully considering skin risk to patient and the risk of removing/replacing the sling for subsequent moves.

---

* "Stand by for safety." In most cases, if a bariatric patient is about to fall, there is very little that the caregiver can do to prevent the fall. The caregiver should be prepared to move any items out of the way that could cause injury, try to protect the patient's head from striking any objects or the floor and seek assistance as needed once the person has fallen.
- If patient has partial weight-bearing capability, transfer toward stronger side.
- Consider using an abdominal binder if the patient's abdomen impairs a patient handling task.
- Assure equipment used meets weight requirements. Standard equipment is generally limited to 250-350 lbs. Facilities should apply a sticker to all bariatric equipment with "EC" (for expanded capacity) and a space for the manufacturer's rated weight capacity for that particular equipment model.
- Identify a leader when performing tasks with multiple caregivers. This will assure that the task is synchronized for increased safety of the healthcare provider and the patient.
- During any patient transferring task, if any caregiver is required to lift more than 35 lbs of a patient's weight, then the patient should be considered to be fully dependent and assistive devices should be used. (Waters, T. [2007]. When is it safe to manually lift a patient? *American Journal of Nursing, 107*[8], 53-59.)

Bariatric Algorithm 5: Patient-Handling Tasks Requiring Access to Body Parts
(Limb, Abdominal Mass, Gluteal Area)
rev. 10/01/08

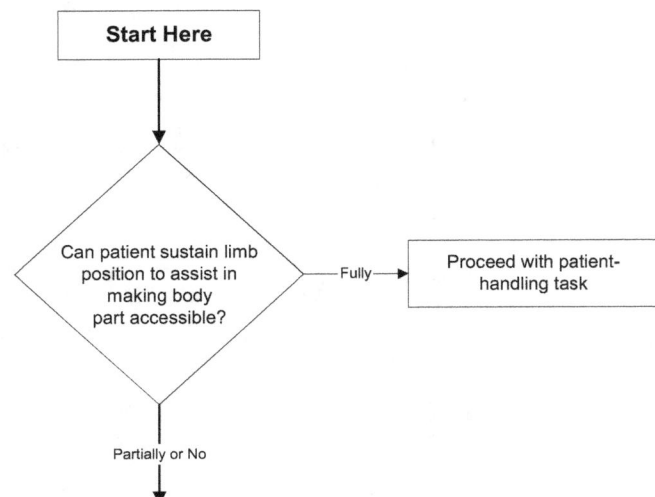

Assemble multidisciplinary team to develop creative solutions that are safe for patient and caregiver.

Examples:
- Modify use of a full body sling lift to elevate limbs for bathing or wound care (i.e. bariatric limb sling).
- Use draw sheet with handles for 2 caregivers (one per side) to elevate abdominal mass to access the perineal area (e.g., catheterization, wound care).
- To facilitate drying a patient between skin folds, use the air-assisted lateral transfer aid to blow air or use a hair dryer on a cool setting.
- Use sealed high-density foam wedge to firmly reposition patient on side. Skid-resistant texture materials vary and come in set shapes and cut-your-own rolls. Examples include:
  - Dycem(TM)
  - Scoot-Guard(TM): antimicrobial; clean with soap and water, air dry.
  - Posey-Grip(TM): Posey-Grip does not hold when wet. Washable, reusable, air dry.

- A multidisciplinary team needs to problem solve these tasks, communicate to all caregivers, refine as needed and perform consistently.
- Consider using an abdominal binder if the patient's abdomen impairs a patient handling task.
- During any patient transferring task, if any caregiver is required to lift more than 35 lbs of a patient's weight, then the patient should be considered to be fully dependent and assistive devices should be used. (Waters, T. [2007]. When is it safe to manually lift a patient? *American Journal of Nursing, 107*[8], 53-59.)

Bariatric Algorithm 6: Bariatric Transporting (Stretcher)
rev. 10/1/08

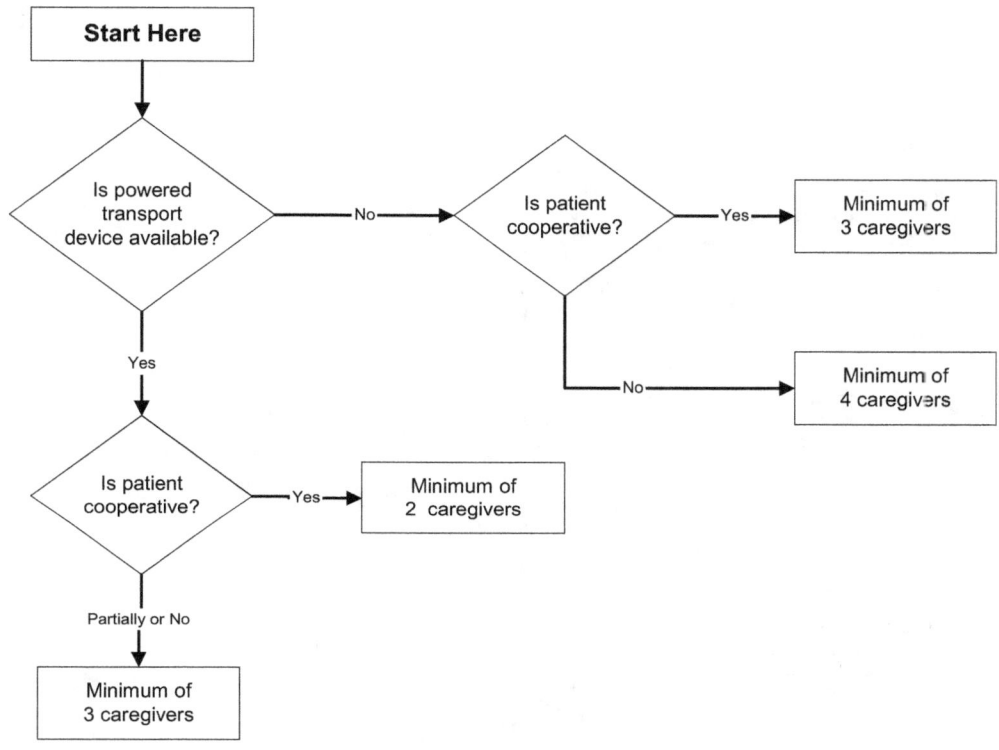

- If the patient has respiratory distress, the stretcher must have the capability of maintaining a high Fowler's position.
- Newer equipment often is easier to propel.
- If patient is uncooperative, secure patient in stretcher.
- During any patient transferring task, if any caregiver is required to lift more than 35 lbs of a patient's weight, then the patient should be considered to be fully dependent and assistive devices should be used. (Waters, T. [2007]. When is it safe to manually lift a patient? *American Journal of Nursing, 107*[8], 53-59.)

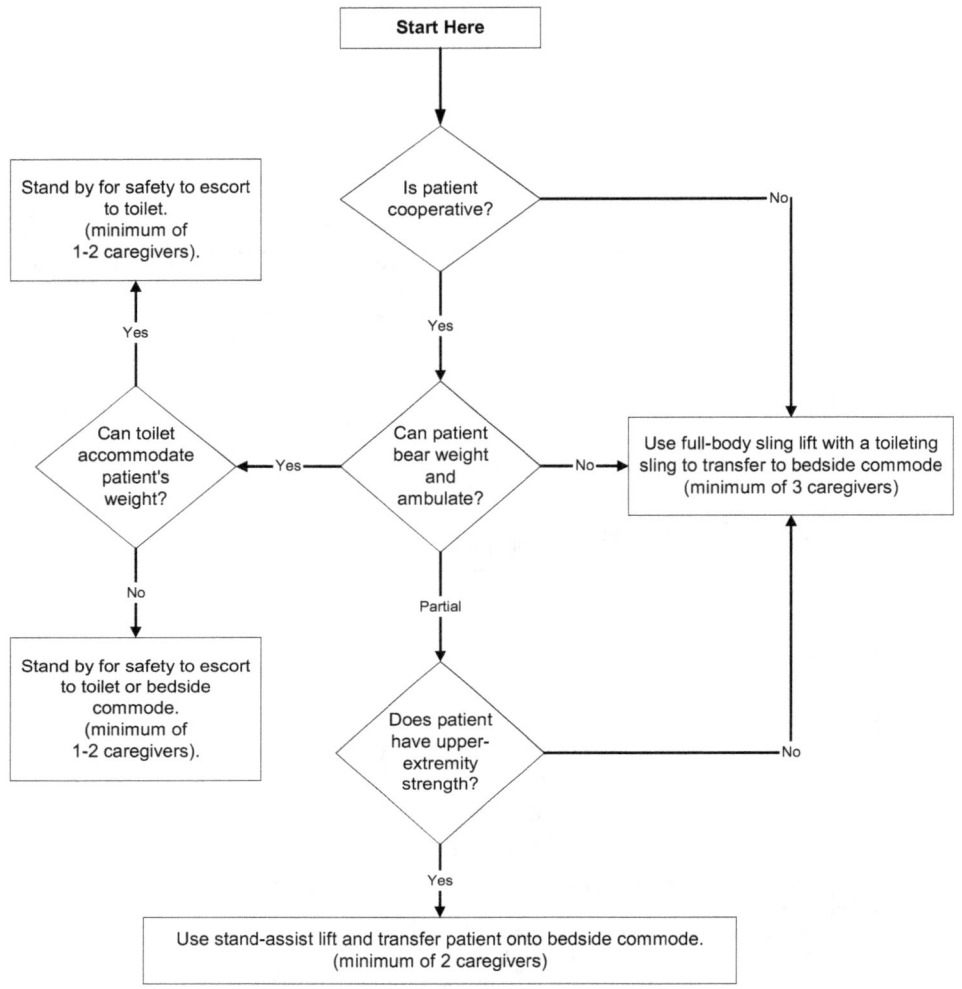

Bariatric Algorithm 7: Toileting Tasks for the Bariatric Patient
rev. 10/01/098

Considerations:
- Is bathroom doorway wide enough to accommote entry of mechanical lift device and patient?
- Assure equipment used meets weight requirements and is appropriately sized for patient.
- Typically, standard toilets are rated to 350 lbs maximum capacity.
- During any patient transferring task, if any caregiver is required to lift more than 35 lbs of a patient's weight, then the patient should be considered to be fully dependent and assistive devices should be used.
(Waters, T. [2007]. When is it safe to manually lift a patient? *American Journal of Nursing, 107*[8], 53-59.)

Bariatric Algorithm 8: Transfer a Bariatric Patient Up From the Floor
Last rev. 10/1/08

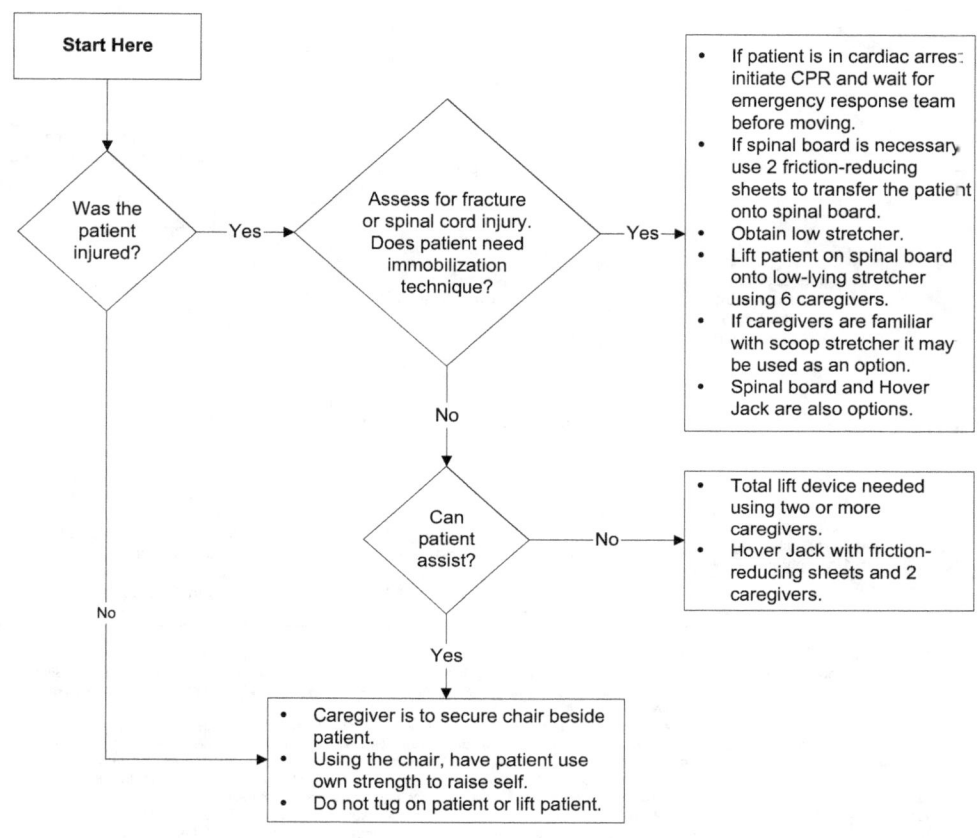

- Do <u>not</u> lift patient off floor.
- Do not allow patient to lean on caregiver for base of support.
- "Immobilization Technique" definition: use spinal precautions if can't use lift due to suspect hip, pelvic, or vertebral fractures.
- Use floor-based lift that goes all the way down to the floor (most of the newer models are capable of this).
- During any patient transferring task, if any caregiver is required to lift more than 35 lbs of a patient's weight then the patient should be considered to be fully dependent and assistive devices should be used. (Waters, T. [2007]. When is it safe to manually lift a patient? *American Journal of Nursing, 107*[8], 53-59.)

Orthopaedic Algorithm #1: Turning Patient in Bed (Side-to-Side)
Patient with Orthopaedic Impairments
September 25, 2008

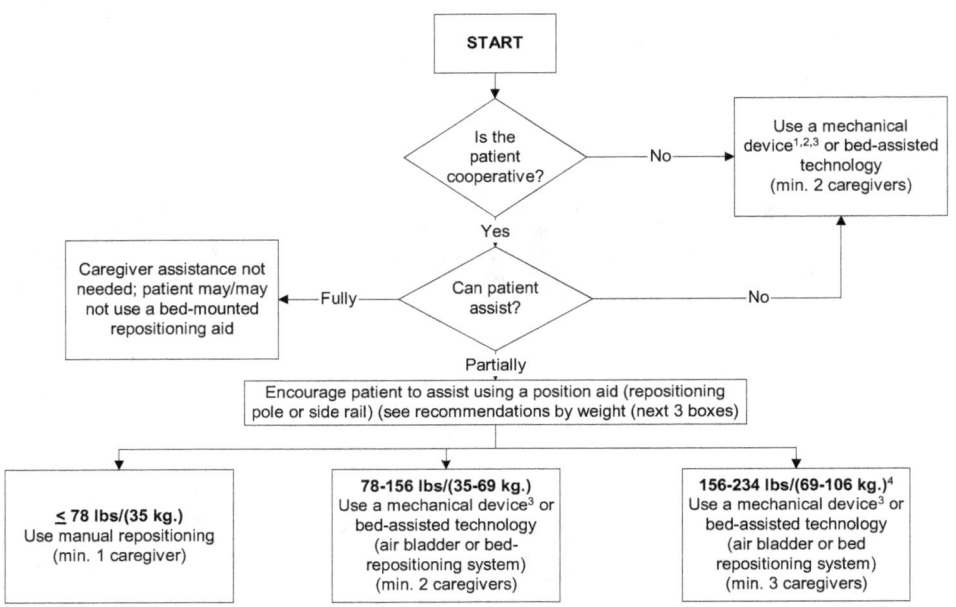

FOOTNOTES:
1. Maintain orthopaedic precautions as prescribed while performing this activity such as total hip, knee, shoulder, or spine precautions.
2. Select sling to meet and maintain the patient's pre-op or post-op positioning guideline/precautions for the affected limb/body part(s). For more information on sling section, see Appendix A.
3. Examples of repositioning mechanical devices are: **Turning clips:** these simple slips attach to a bed sheet and can be used with a floor-based lift or ceiling-based lift to facilitate turning a patient. **Turning straps/slings:** one end of these straps or slings is connected to the bed and the other end is attached to either a ceiling or floor based lift to facilitate turning the patient. **Powered mechanical devices:** a ceiling lift is a powered overhead lift that can be used with a repositioning sling to turn a patient in bed. **Friction reducing devices:** either tubular in design, or two separate pieces of material are placed under the patient to assist in turning the patient in bed or moving the patient to the head of the bed. **Pulley systems:** these devices work by use of a pulley system and an overhead frame. The user turns a crank, which engages the pulley system to retract straps that are connected to a rod and bed sheet, thus turning the patient on the side.
4. If the patient weighs more than 234 lbs. mechanical assistive devices should be used to assist. Use your best clinical judgment for the number of caregivers required to assist.

GENERAL NOTES:
- For any patient who has, or is at risk for a pressure ulcer, care should be taken to avoid shearing force (such as using a friction reducing device for repositioning in bed). Shearing force is when there are two forces moving in opposite directions adjacent to each other (like scissors).
- The height of the bed should be appropriate for staff safety (at elbow height).
- During any patient handling task, if the caregiver is required to lift more than 35 lbs./(16 kg.) of a patient's weight, then the patient should be considered fully dependent and an assistive device should be used. (Waters, T. [2007]. When is it safe to manually lift a patient? *American Journal of Nursing, 107*(8), 53-59).

Orthopaedic Algorithm #2: Vertical Transfer of a Post-Operative Total Hip Replacement Patient
(Bed to Chair, Chair to Toilet, Chair to Chair, or Car to Chair)
September 25, 2008

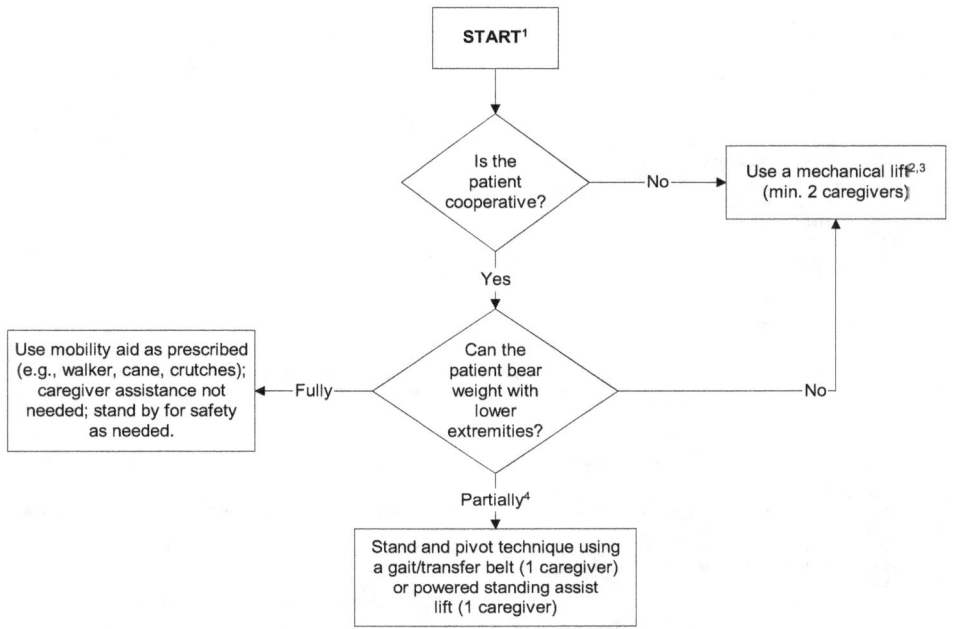

FOOTNOTES:
1. See 1A, 1B, 1C, 1D below for techniques to position patient at side of bed.
    1A. Moving from supine head of bed elevated to sitting at edge of bed requires: Patient's ability to shift their seated weight in a sitting position. Typically accomplished by unweighting one buttock and moving it toward the edge of the bed; repeating this in alternating fashion until patient is sitting at edge of bed.
    1B. With an impaired upper or lower extremity, caregiver might need to support the limb while patient attempts #1A.
    1C. If patient is unable to accomplish #1A with #1B and the amount of assistance from caregiver will exceed 35 lbs., then a mechanlical lift device should be used to achive sitting position at the edge of the bed.
    1D. Anti-friction sheets and seated discs might be useful when the amount of caregiver assistanace is close to recommended limits; be aware of skin shearing risks. Shearing forces are caused when there are two forces moving in opposite directions adjacent to each other (like scissors).
2. Maintain orthopaedic precautions as prescribed while performing this activity such as total hip, knee, shoulder, or spine precautions.
3. Select sling to meet and maintain the patient's pre-op or post-op positioning guideline/precautions for the affected limb/body part(s). For more information on sling section, see Appendix A.
4. This will include situations where the patient may be allowed: a) Limited weight bearing on one lower extremity and full weight bearing on the other extremity; b) Partial weight bearing through both lower extremities.

GENERAL NOTES:
- If patient has partial weight bearing capacity, transfer toward stronger side.
- For car transfers: a) If patient cannot tolerate a seated position when doing a car transfer use a stretcher transfer or alternative transportation may be required; b) All car transports should comply with state laws for both children and adults; c) Don't forget to use all of the features of the car (ie., adjustability of the seat) during the transfer.
- The height of the bed should be appropriate for staff safety (at elbow height).
- During any patient handling task, if the caregiver is required to lift more than 35 lbs./(16 kg.) of a patient's weight, then the patient should be considered fully dependent and an assistive device should be used. (Waters, T. [2007]. When is it safe to manually lift a patient? *American Journal of Nursing*, 107(8), 53-59).

Orthopaedic Algorithm #3: Vertical Transfer of a Patient with an Extremity Cast/Splint
September 25, 2008

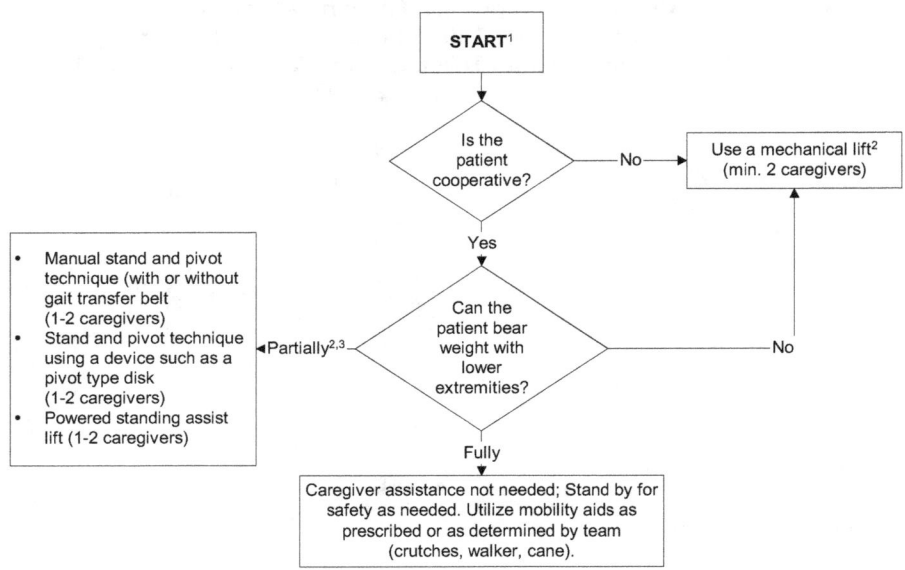

FOOTNOTES:
1. Moving from supine head of bed elevated to sitting at edge of bed requires a patient's ability to shift their seated weight in a sitting position:
a. When assistance is not required, this is typically accomplished by unweighting one buttock and moving it toward the edge of the bed; repeating this in alternating fashion, until patient is sitting at the edge of the bed.
b. With an impaired upper or lower extremity:
- if the amount of assistance from caregiver does not exceed 35 lbs., caregiver may provide limb support while patient moves unassisted to side of bed (see a. above)
- if the amount of assistance from caregiver may exceed 35 lbs., then a limb support strap/sling with a mechanical lift will provide limb support while patient moves unassisted to side of bed (see 1a. above)
c. If patient is unable to accomplish a. and/or b. then utilize one of the following options:
- mechanical lift device with a seated sling to lift patient to side of bed
- friction-reducing device to assist staff in pulling patient to side of bed
d. Friction-reducing devices and seated discs may be useful when the amount of caregiver assistance is close to recommended limits, but be aware of skin shearing risks. Shearing is caused when there are two forces moving in opposite directions adjacent to each other (like scissors).
2. Select sling to meet and maintain the patient's pre-op or post-op positioning guideline/precautions for the affected limb/body part(s). For more information on sling selection, see Appendix A.
3. Patient can bear weight on one leg only (e.g., weight bearing on unaffected limb or limited weight bearing on affected limb).

GENERAL NOTES:
- Need to test the fit of the sling with an immobilized extremity.
- Maintain affected extremity immobilization/alignment.
- Use lift device with limb sling if applicable.
- During any patient handling task, if the caregiver is required to lift more than 35 lbs./(16 kg.) of a patient's weight, then the patient should be considered fully dependent and an assistive device should be used. (Waters, T. [2007]. When is it safe to manually lift a patient? *American Journal of Nursing*, 107(8), 53-59).

Orthopaedic Algorithm #4: Ambulation
September 25, 2008

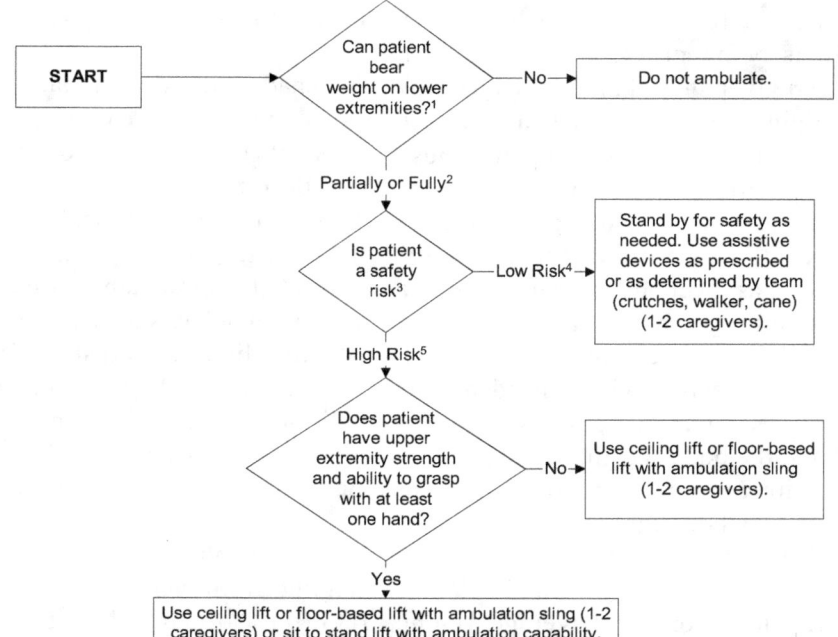

FOOTNOTES:
1. Non-weight bearing: Patient is unable to bear weight through both lower extremities or weight-bearing through both lower extremities is contraindicated.
2. Partial weight bearing: This will include situations where the patient may be allowed: a) Limited weight bearing on one lower extremity and full weight bearing on the other extremity; b) Partial weight bearing through both lower extremities.
3. Safety risks may include: decreased cognition; decreased ability to cooperate/ combativeness; medical stability.
4. Factors that contribute to low safety risk: a) Lack of combativeness; b) Ability to follow commands; c) Medical stability; d) Experience with the assistive device.
5. Factors that contribute to high safety risk: a) Combativeness; b) Lack of ability to follow commands; c) Medical instability; d) Lack of experience with the assistive device, e) neurological deficits.

GENERAL COMMENTS/DISCUSSION:
- In healthcare, weight-bearing is often used to describe the amount of weight bearing that the patient can or has done. In orthopedics, weight-bearing status is prescribed by the physician based on the patient's ability to safely bear weight through the musculoskeletal system. Exceeding the prescribed weight-bearing status may be detrimental to the patient.
- Patients should be assessed for safety risks as described above. If patients are determined to be at significant risk for falls, then care givers assisting with ambulation are also at risk for assisting patients to prevent fall. In high risk situations precautions should be taken, and devices such as walking slings should be used. At some point in care, the team will need to weigh the risks of falls with the benefits of ambulation and take a "therapeutic" risk in order to functionally advance the patient.
- Need to test the fit of the sling with an immobilized leg. For more information on on sling selection, see Appendix A.
- Maintain affected extremity immobilization/alignment.
- During any patient handling task, if the caregiver is required to lift more than 35 lbs./(16 kg.) of a patient's weight, then the patient should be considered fully dependent and an assistive device should be used. (Waters, T. [2007]. When is it safe to manually lift a patient? *American Journal of Nursing*, 107(8), 53-59).

Orthopaedic Clinical Tool #1:
Lifting and Holding Legs or Arms in an Orthopaedic Setting

Introduction

Often when giving orthopaedic care, the caregiver must lift and/or hold a limb in place while some type of treatment is being provided, such as cast application. The caregiver should maintain a neutral body posture (upright, not fully flexed) and adjust the height of the table. When a caregiver must lift a patient's leg or arm, it is important to make sure that the weight of the limb being lifted does not exceed the strength capability of the caregiver. An ergonomic tool has been developed to assist caregivers in determining whether a specific lift and/or hold of a limb is acceptable and whether some type of lift or hold assist device is needed. The tool, which is presented in Table 1, is based on the limb being lifted and the body weight of the patient. For lifts of limbs with casts, an alternate method is presented for assessing whether the lift is acceptable or not as presented in Table 2.1.

This tool shows the calculation of the average weight for an adult patient's leg and arm as a function of whole body mass. Weights are presented both in pounds (lbs.) and metric (kg) units. Maximum lift and hold loads were calculated based on 75th percentile shoulder flexion strength and endurance capability for U.S. adult females, where the maximum weight for a one-handed lift is 11.1 lbs. and for a two-handed lift is 22.2 lbs. [Waters et al. 2009].

The shaded areas of the table indicate whether it is acceptable for one caregiver to lift the listed body parts with one or two hands or hold the body parts for 1, 2, or 3 minutes with two hands. Respecting these limits will minimize risk of muscle fatigue and the potential for MSDs. If the limb weight exceeds the values listed in the table, use of assistive technology is recommended, such as a ceiling lift or floor-based lift with a limb support sling. Orthopaedic caregivers must use clinical judgment to assess the need for additional staff member assistance or assistive devices to lift and/or hold one of these body parts for a particular period of time.

Patient weight is divided into weight categories ranging from very light to morbidly obese (see Table 1). Normalized weight for each leg and each arm are calculated as a percentage of body weight where each complete arm weighs 5.1% of total body mass and each leg weighs 15.7% of total body mass [Chaffin et al. 1999]. All weights are presented in both pounds and kilograms, rounded to the nearest whole unit.

References for Clinical Tool #1

Chaffin D B, Anderson GBJ, Martin BJ [1999]. Occupational biomechanics. 3rd ed. New York: J. Wiley and Sons.

Pheasant S [1992]. Bodyspace. London, England: Taylor and Francis.

Rohmert W [1973a]. Problems of determination of rest allowances. Part 1: Use of modern methods to evaluate stress and strain in static muscular work. Appl Ergon 4(2):91–95.

Rohmert W [1973b] Problems of determination of rest allowances. Part 2: Determining rest allowances in different human tasks. Appl Ergon 4(3):158–162.

Waters T [2007]. When is it safe to manually lift a patient? Am J Nurs 107(8):53–59.

Waters T, Sedlak C, Howe C, Gonzalez C, Doheny M, Patterson M, and Nelson A. [2009] Recommended Weight Limits for Lifting and Holding Limbs in the Orthopaedic Practice Setting. Orthopaedic Nursing, Vol 28(2S), S28-S32.

Table 1.1 Ergonomic tool: Lifting and holding legs or arms in an orthopaedic setting*

Patient Weight lbs. (kg.)	Body Part	Body Part Weight Lbs. (kg.)		Lift 1-hand	Lift 2-hand	Hold 2-hand 1 min.	Hold 2-hand 2 min.	Hold 2-hand 3 min.
<40 lbs. (< 18 kg.)	Leg	<6.3 lbs.	(3 kg)					
	Arm	<2.0 lbs.	(1 kg)					
40–90 lbs. (18–41 kg)	Leg	<14.1 lbs.	(6 kg)	grey				grey
	Arm	<4.6 lbs.	(2 kg)					
90–140 lbs. (41–64 kg)	Leg	<22.0 lbs.	(10 kg)	grey			grey	grey
	Arm	<7.1 lbs.	(3 kg)					
140–190 lbs. (64–86 kg)	Leg	<29.8 lbs.	(14 kg)	grey	grey		grey	grey
	Arm	<9.7 lbs.	(4 kg)					
190–240 lbs. (86–109 kg)	Leg	<37.7 lbs.	(17 kg)	grey	grey	grey	grey	grey
	Arm	<12.2 lbs.	(6 kg)					
240–290 lbs. (109–132 kg)	Leg	<45.5 lbs.	(21 kg)	grey	grey	grey	grey	grey
	Arm	<14.8 lbs.	(7 kg)					grey
290–340 lbs. (132–155 kg)	Leg	<53.4 lbs.	(24 kg)	grey	grey	grey	grey	grey
	Arm	<17.3 lbs.	(8 kg)				grey	grey
340–390 lbs. (155–177 kg)	Leg	<61.2 lbs.	(28 kg)	grey	grey	grey	grey	grey
	Arm	<19.9 lbs.	(9 kg)				grey	grey
390–440 lbs. (177–200 kg)	Leg	<69.1 lbs.	(31 kg)	grey	grey	grey	grey	grey
	Arm	<22.2 lbs.	(10 kg)			grey	grey	grey
> 440 lbs. (>200 kg)	Leg	>69.1 lbs.	(31 kg)	grey	grey	grey	grey	grey
	Arm	>22.2 lbs.	(10 kg)	grey	grey	grey	grey	grey

* **No shading**: Lift and hold is appropriate but use clinical judgment and do not hold longer than noted.
Grey shading: Do not lift alone; use assistive device or more than one caregiver.
From Waters et al., 2009.

Note: It is important to remember that the table shows the acceptable weights for limbs without a cast in place. If the caregiver is lifting a limb with a cast, the additional weight of the cast should be added to the weight of the limb to determine whether the lift is acceptable. An alternate method is provided below for limbs with casts. These are guidelines for the average weight of the leg and arm and are presented in Table 2.1.

Rationale for Table 1.
To accommodate 75% of the U.S. adult female working population, maximum load for a one-handed lift is calculated to be 11.1 lbs. (5.0 kg). This is determined by calculating the strength capabilities for 25th percentile U.S. adult female maximum shoulder flexion movement (the mean equals 40 Newton meters [Nm], standard deviation equals 13 Nm) [Chaffin et al. 1999] and 75th percentile U.S. adult female shoulder to grip length (the mean equals 610 mm, the standard deviation equals 30 mm) [Pheasant 1992]. Maximum loads for one person for a 2-handed lift (i.e., 22.2 lbs./10.1 kg.) are calculated as twice that of a one-handed lift. Muscle strength capabilities diminish as a function of time; therefore, maximum loads for 2-handed holding of body parts are presented for 1, 2, and 3 minute durations. After 1 minute, muscle endurance has decreased by 48%, decreased by 65% after 2 minutes, and, after 3 minutes of continuous holding, strength capability is only 29% of initial lifting strength [Rohmert 1973a; 1973b]. If the limits in ergonomic Table 1 are exceeded, additional staff members or assistive limb holders should be used.

Orthopaedic Clinical Tool #2: Alternate Method for Determining Safe Lifting and Holding of Limbs with Casts

Table 2.1. Limb weight factor for lifting or holding a limb with a cast.

Limb	Limb Weight Factor	Lift 1-hand	Lift 2-hands	Hold 2-hands 1 min.	Hold 2-hands 2 min.	Hold 2-hands 3 min.
Leg	0.157	11.1 lbs. (5.1 kg)	22.2 lbs. (10.2 kg)	11.6 lbs. (5.3 kg)	7.8 lbs. (3.5 kg)	6.4 lbs. (2.9 kg)
Arm	0.051					

From Waters et al., 2009.

To determine whether it is acceptable to lift or hold a limb with a cast, multiply the patient's weight times the limb factor (0.157 for leg and 0.051 for arm) and add the weight of the cast to obtain the total limb weight. Compare the calculated total limb weight to the value in the appropriate task box (e.g. 11.1 lbs for 1 handed lift, 22.2 lbs for 2 handed lift, etc.). If the total limb weight exceeds the weight in the appropriate box, then the caregiver should not manually lift the limb alone but should use an assistive device or more than one caregiver to perform the lift. On the other hand, if the calculated weight is less than the value in the appropriate box, then it is acceptable to manually lift and hold the limb, and the caregiver should use clinical judgment and not hold longer than noted.

As an example, if the patient weighs 200 lbs. and has an arm cast weighing 5 lbs., then the total arm weight would be 200 lbs. × 0.051 plus 5 lbs., or 15.2 lbs. In this case, the arm should not be lifted with one hand (i.e., 15.2 lbs. > 11.1 lbs.). It could be lifted with two hands (i.e., 15.2 lbs. < 22.2 lbs.) but not held in that position for more than a few seconds (15.2 lbs. > 11.6 lbs.). Similarly, if the patient weighs 75 lbs. and has a 5 lb. leg cast, then the total limb weight would be 75 lbs. x 0.157 plus 5 lbs., or 16.8 lbs. In this case, it would be unacceptable to lift the limb with one hand (i.e., 16.8 lbs. > 11.1 lbs.). However, lifting with two hand would be acceptable (i.e., 16.8 lbs. < 22.1 lbs.), but the limb should not be held for more than a few seconds (16.8 lbs. > 11.6 lbs.).

Table 2.2. Predicted Weights for a Fiberglass Cast

The following Table 2.2 provides some predicted weights for a fiberglass cast.

Short Arm Cast (adult)	Long Arm Cast (adult)	Short Leg Walking Cast (150 lbs. adult)	Long Leg Cast (150 lbs. adult)	Infant Body Spica 20-30 lbs.	Child Body Spica 3-5 yr old 30-50 lbs.
0.5 lbs.	1 lbs.	2 lbs.	3.0 lbs.	2 lbs.	4 lbs.
2 rolls 3"	1 roll 2" 3 rolls 3"	4 rolls 4"	3 rolls 3" 3 rolls 4"	2 rolls 2" 3 rolls 3"	5 rolls 3" 5 rolls 4"
+ webril*	+ webril*	+ webril*	+ webril*	+ webril*	+ webril*

*Weight of webril is 0.25 lb. per packet
From Waters et al., 2009.

Appendix B - Assessment Criteria and Care Plan for Safe Patient Handling and Movement

I. **Patient's Level of Assistance:**
   _____ Independent— Patient performs task safely, with or without staff assistance, with or without assistive devices.
   _____ Partial Assist— Patient requires no more help than standby, cueing, or coaxing, or caregiver is required to lift no more than 35 lbs of a patient's weight.
   _____ Dependent— Patient requires nurse to lift more than 35 lbs of the patient's weight, or patient is unpredictable in the amount of assistance offered. In this case assistive devices should be used.

An assessment should be made prior to each task if the patient has varying level of ability to assist due to medical reasons, fatigue, medications, etc. When in doubt, assume the patient cannot assist with the transfer/repositioning.

II. **Weight-Bearing Capability**
   _____ Full
   _____ Partial
   _____ None

III. **Bilateral Upper-Extremity Strength**
   _____ Yes
   _____ No

IV. **Patient's level of cooperation and comprehension:**
   _____ Cooperative—may need prompting; able to follow simple commands.
   _____ Unpredictable or varies (patient whose behavior changes frequently should be considered as unpredictable), not cooperative, or unable to follow simple commands.

V. **Weight:** _____ **Height:** _____
   Body Mass Index (BMI) [needed if patient's weight is over 300 lbs][1]:
   If BMI exceeds 50, institute Bariatric Algorithms

The presence of the following conditions are likely to affect the transfer/repositioning process and should be considered when identifying equipment and technique needed to move the patient.

VI. **Check applicable conditions likely to affect transfer/repositioning techniques.**

_____ Hip/Knee/Shoulder Replacements	_____ Respiratory/Cardiac Compromise	_____ Fractures
_____ History of Falls	_____ Wounds Affecting Transfer/Positioning	_____ Splints/Traction
_____ Paralysis/Paresis	_____ Amputation	_____ Severe Osteoporosis
_____ Unstable Spine	_____ Urinary/Fecal Stoma	_____ Severe Pain/Discomfort
_____ Severe Edema	_____ Contractures/Spasms	_____ Postural Hypotension
_____ Very Fragile Skin	_____ Tubes (IV, Chest, etc.)	

Comments:_____
_____
_____
_____
_____

VII. Appropriate Lift/Transfer Devices Needed:

Vertical Lift:

Horizontal Lift:

Other Patient-handling Devices Needed:

Sling Type Seated_____ Seated (Amputee) _____ Standing_____ Supine_____ Ambulation_____ Limb Support_____

Sling Size _____

Signature _____ **Date** _____

[1]If patient weighs more than 300 lbs, the BMI is needed. For Online BMI table and calculator see: http://www.nhlbi.nih.gov/guidelines/obesity/bmi_tbl.htm

www.ingramcontent.com/pod-product-compliance
Lightning Source LLC
Chambersburg PA
CBHW081805170526
45167CB00008B/3340